In Focus
Vic Berry's

Ashley Butlin

Coorlea Publishing,
West Lodge,
Taverham Hall,
Norwich, NR8 6HU

ISBN 0 948069 07 4

Design: Ashley Butlin

Editor: Jenny Butlin

Typesetting: PTPS, 4 Taverners Sq., Norwich

Printers: Haynes Cannon Ltd., Brunel Close,
Wellingborough, Northants.

Acknowledgements

In compiling this record of Vic Berry's, my grateful thanks are due to a number of people for their invaluable help and assistance. David Pearce and David Webb have given essential historical information. My principal photographers, Paul Biggs and Richard Lillie, have sacrificed many hours of their time and must have despaired with several of my requests. My good friend and colleague Peter Beer has been extremely helpful with advice. Mr.Vic Berry, along with his management and staff, has given every encouragement, and without their full co-operation, this publication would not have been possible. Finally, I must thank my wife Jenny for her total involvement and help in the production of this book.

Front cover: ***Cutting the Class 20s, November 1987.*** *Photo: Richard Lillie.*

Back cover: ***A typical scene from the yard during the Autumn of 1987.*** *Photo: Richard Lillie.*

Above: ***Sparks fly as yet another locomotive is reduced to manageable pieces of scrap metal.*** *Photo: Richard Lillie.*

Title page: ***The demise of yet another Class 27.*** *Photo: Roger Bamber.*

Inside covers: Photos: Roger Bamber.

Foreword

Little did I imagine when, as a six-year-old lad, my Grandfather used to take me to watch the trains from Upperton Road bridge working in Braunstone Gate Goods Yard, that one day I would run a major railway scrapyard and rolling stock refurbishing business from that site.

Although I would not describe myself as a dedicated enthusiast, I have always taken an interest in railways. However, it was not until I set up in business in 1973 and the first consignment of wagons arrived for cutting, that I realised a business could also be a hobby. Now some 15 years later, that is still very much the case.

I must admit, I remain slightly mystified by the fanatical obsession some enthusiasts have for the yard, although I can appreciate their interest. One thing I can promise: 1988 at Vic Berry's will be as interesting for enthusiasts as it will be, in business terms, for me.

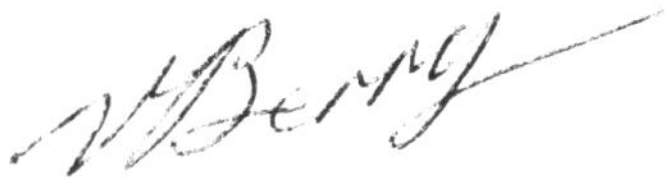

Vic Berry.

The man who has become a household name amongst enthusiasts, Mr. Vic Berry, stands in front of the famous stack, contemplating his next acquisition. Photo: Richard Lillie.

Introduction

In July 1898, freight traffic began to run on the newly built Great Central Railway, in advance of passenger traffic. These early freight workings were designed to help to consolidate the track bed ready for passenger workings which commenced some nine months later. With the introduction of freight, a goods yard was opened in Leicester at Braunstone Gate. It was situated to the south of Leicester Central station on land to the east of the main line. Prominent in the yard was the large goods depot building which remains today in use as a warehouse, a monument to Victorian architecture.

The fortunes of the Great Central are well known. With its route to London from the north mirroring existing lines, it never realised its full potential. Ironically, the last mainline to London was also the first to close. However, in the early sixties, prior to closure, freight traffic to Braunstone Gate was reported to have served three adjoining industries: two scrap merchants despatching metal to steelworks, blast furnaces and foundries as far afield as Sheffield, Scunthorpe and South Wales, and a timber merchant who used BR to transport imports from both East and West coast ports to its private siding.

A view from Upperton Road bridge in 1964 of Braunstone Gate Goods Yard. Prominent in the centre of the yard is the large goods depot built by the Great Central Railway in the 1890s. To the right are the brick buildings of the former Electric Light Works. To the left of the photograph, the four tracks are the mainline and goods loops, now part of the Great Central Way footpath. Beyond the signal box, the corrugated carriage sheds have been demolished but the brick built offices at the side of them remain. Ironically, the crippled mineral wagon in the foreground stands in the exchange sidings, used today to store stock awaiting entry to the yard.
Photo: David Webb.

Such was the value of this traffic to BR that a connecting spur was built on land occupied by the Leicester Central locomotive sheds, from the Leicester/Burton-on-Trent line to the Great Central just south of Braunstone Gate. This spur was formally opened on 14th.June 1965 at which point through-freight workings on the GC came to an end, although passenger traffic continued until May 1969. Braunstone Gate was now served by trip workings from freight yards on the Midland line, a situation which remains today. During the early sixties, Leicester Midland TMD was home to a number of Class 25s and 27s, and these were often employed on trips to Braunstone Gate. The official BR photograph taken of the opening of the new spur, published in the September 1965 "Modern Railways", shows an unidentified Class 27 at the head of a mixed freight. Twenty years later, 25s and 27s were again on trip workings to Braunstone Gate but these occasions hauled for their final journey.

It was against a tradition of scrap dealers at Braunstone Gâte that Mr. Vic Berry began his entrepreneurial operations in 1973. His first consignments of stock from BR consisted of redundant mineral wagons. So many arrived that they were stacked five high in his yard. From these beginnings, Vic Berry moved on to tackle hauled carriage stock and redundant 4-SUB trailers. By the end of the seventies, the name of Vic Berry was becoming well known amongst enthusiasts who followed the fortunes of withdrawn stock, although it was to be 1983, when three Class 76 electrics arrived, before reports of his yard began to appear in the popular monthly magazines. When, in 1986, locomotives appeared in vast numbers, interest in the yard escalated and Vic Berry's became a household name to enthusiasts throughout the country. During 1987, with the building of the now famous stack, Vic Berry's have been approached by numerous people wishing to visit the yard. Many of these are enthusiasts just hoping to have a last look at stock.

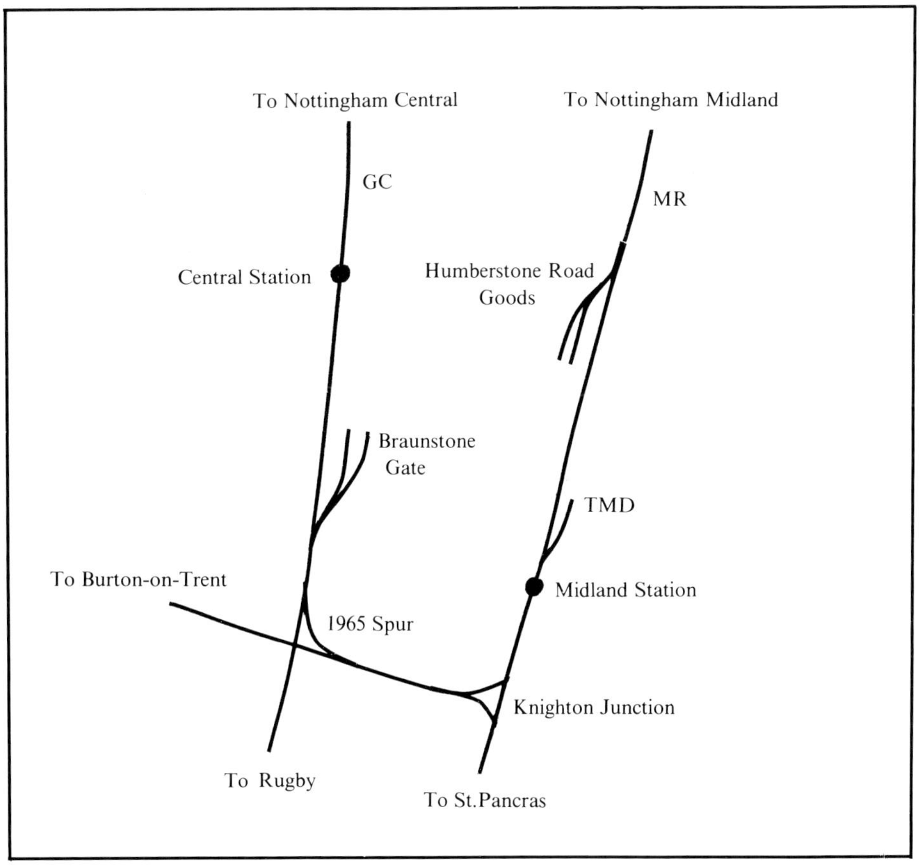

Above: ***Map showing relative positions of the lines and locations in Leicester referred to within the text.***

Below: ***Braunstone Gate Goods Yard circa. 1905. Comparison with the photograph opposite shows how further tracks leading to the Goods Shed were added at a later date. In addition, the freight loops between the yard and main line were later extended.***

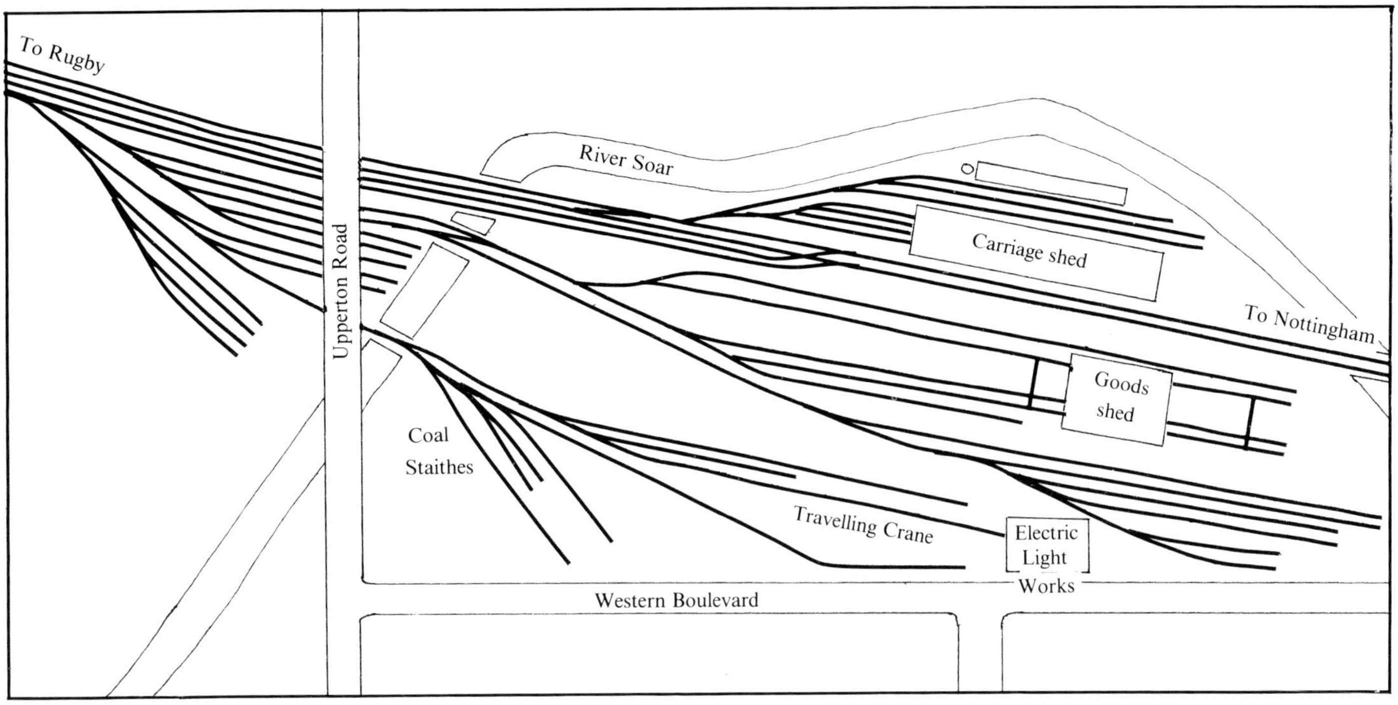

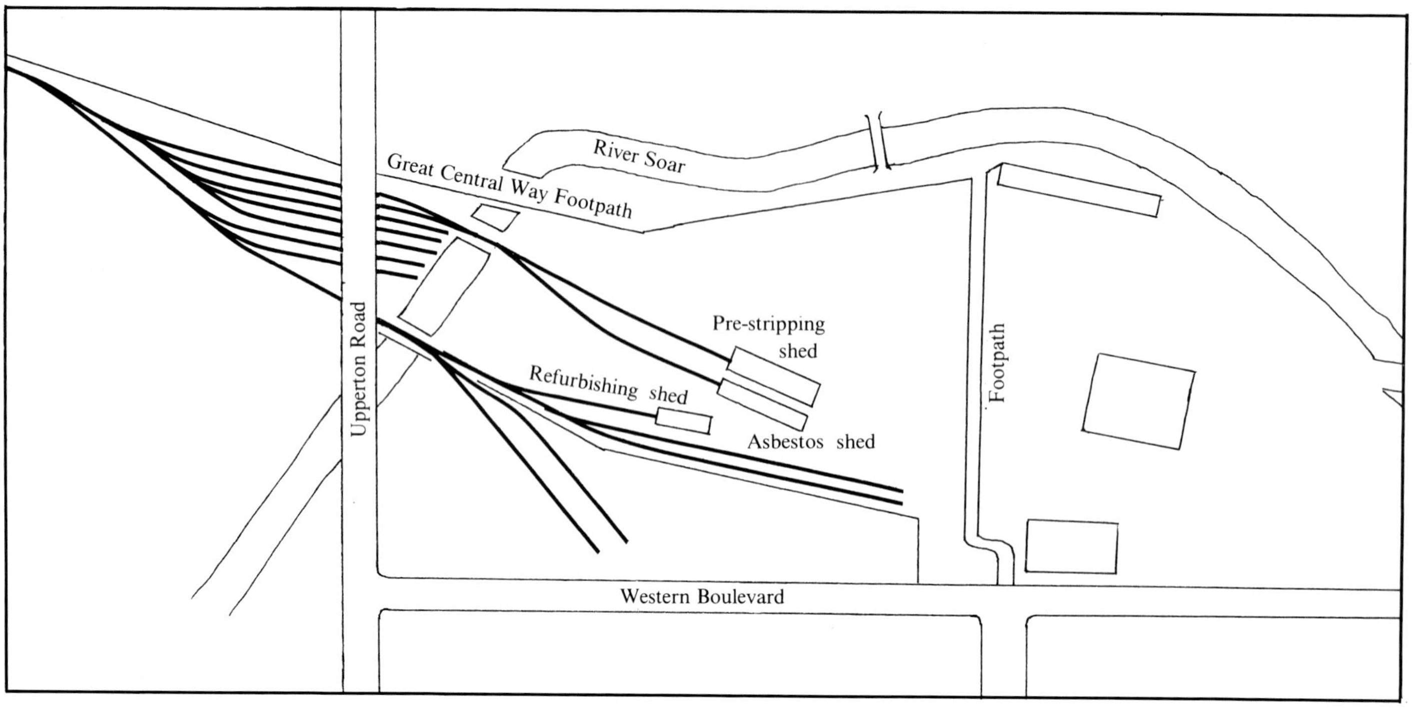

Above: Braunstone Gate 1987. Only the southern end of the former goods yard has been taken over by Vic Berry's, the northern end being occupied by other industrial enterprises.

Preservation societies have been able to obtain not only complete locomotives but also numerous components. Less expected were the television crews who saw the stack as a newsworthy item. As a result, it was featured several times during that Summer. The yard has also acted as a location for photographic assignments. Not foreseen was a request from a theatre group for props for use in a local stage production.

Stock destined for Vic Berry's arrives at Leicester Humberstone Road sidings which are located to the west of the Midland mainline and north of Leicester station. From here stock is tripped to Braunstone Gate as space at the yard allows. At Braunstone Gate, five sidings located under Upperton Road bridge are used as exchange sidings where stock is stored prior to entering the yard. Vic Berry's former BR Class 03 shunter 03069 is then used to move stock into the yard as required. It has been a shortage of storage space at Braunstone Gate that has resulted in the building of the various stacks at the yard. Not only have Vic Berry's cut stock at their Leicester yard, but they were also involved in the cutting of the last few locomotives remaining at Swindon in 1986, following the closure of the works there. During the first few months of 1987, the firm cut a number of locomotives at Thornton Yard in Scotland, and at various times, cutters have gone to depots to cut stock which cannot be moved by rail.

Vic Berry's are involved not only in the cutting of stock at Braunstone Gate. British Rail have an undertaking to have no stock containing asbestos in service by the close of 1988. Much has been or will be withdrawn, but this is not always the most financially viable option available. Stock with a good life expectancy can have asbestos removed from within it and the vehicle returned to traffic. Vic Berry's have pioneered this work in recent months and seem set to take on much more of it during 1988.

At the time of writing, numerous withdrawn locomotives, units and carriages litter BR yards throughout the country. Many of these may well be sold to Vic Berry's and the arrival at the yard on the 8th.February, 1988 of the first Class 50, No.50006, came as no real surprise. A new stack composed of Classes 33,45,47 & 50s is certainly feasible. Withdrawals of DMUs continue at a very high rate and their presence at Vic Berry's is certain. With BR's ongoing programme of building new stock, Vic Berry's, the most famous locomotive scrapyard in Britain in the eighties, seems likely to be kept busy for many years disposing of BR's redundant assets.

Locomotives cut by Vic Berry's at depots.

1985

November
45054 Toton

December
25261 Derby Works

1986

February
27203 Etches Park

March
25033 Etches Park
25088* Etches Park

April
45068* Allerton

August
45014* Ashton Road, Manchester

1987

January
26044 Thornton Yard

February
27020 Thornton Yard
27033 Thornton Yard

March
08173 Thornton Yard
26030* Thornton Yard
26033* Thornton Yard
27032* Thornton Yard
27036* Thornton Yard

April
27021* Thornton Yard
27027* Thornton Yard
27029* Thornton Yard

May
26019* Thornton Yard

* cabs returned to Leicester for processing.

Above: ***The present-day view from Upperton Road bridge obviously shows great changes to the yard but also, in the background, many similarities with the earlier view. At the far end of the yard, the GC Goods Shed still dominates the scene while, to the right, the former Electric Light Works buildings remain intact. Perhaps surprisingly, most of the tall factory chimneys still stand. Note the lifebelt hanging from the underside of the upturned cab on the riverside.*** *Photo: Richard Lillie.*

Below: ***During 1986, Vic Berry's became involved in the cutting of locomotives at Swindon Works. As some of the cabs were found to contain asbestos, they were returned to Leicester to allow for its safe removal prior to final cutting. Cabs from Nos.25139 and 25256 await entry into the yard on the 10th. October, 1986.*** *Photo: Paul Biggs.*

Above: ***Not all which leaves the yard is destined to be melted down. A lorry loaded with six sets of wheels from Class 116 units stands on the weighbridge on the 16th.December, 1987, awaiting departure to BRML Doncaster where they will be refurbished for further use.*** *Photo: Richard Lillie.*

Below: ***The arrival at the yard of the first Class 50 No.50006 on the 8th.February 1988 in company with Class 45 No.45144, confirms Mr.Vic Berry's prediction that 1988 will be a year of interest to the enthusiasts.*** *Photo: Richard Lillie.*

Class 20

When, in June 1957, No.D8000 (later No. 20050) emerged from Vulcan Foundry, it was to herald a new era in the history of British railways, for No.D8000 was the first locomotive to be built as a result of the 1955 Modernisation Plan. Thirty years on, D8000 now resides in the NRM at York, although the vast majority of the 228-strong class remain in service. Of the twenty cut for scrap, Glasgow Works has been responsible for twelve, while Crewe and Derby have cut three and two respectively. In November 1987, three of the class, Nos.20155/62/91, resided at BRML Doncaster, stripped of all re-usable components and were officially in a stored status. When it was apparent that components would not be available to repair these three, along with a further ten stored at March, Haymarket and Toton, they were all withdrawn. The Doncaster trio, the first of the class to be sold to a private contractor, were almost immediately lifted off their bogies and dispatched by road to Vic Berry's. All three were placed in front of the stack and cut within days of arrival. With repairs to the class now being of a limited nature, it is most unlikely that these three will be the sole representatives of the class to be sold to Vic Berry's.

Class 20 arrival dates.
1987
November
20155/62/91

Below: All three Class 20s were delivered to Vic Berry's by road during November 1987. Devoid of bogies, they were placed in front of the stack at right angles to each other and were cut within days of arrival. When photgraphed on the 20th. of the month, only No.20191 remained virtually intact, while the demolition of No.20162, Above, was well advanced. Photos: Richard Lillie.

Class 25

The Class 25s were first introduced into traffic in 1961, being a design improvement of the earlier built Class 24s. Over the period 1963-67, 327 were constructed, mostly at Derby, but with sixty-five coming from Darlington Works and thirty-six from Beyer Peacock. When new, the vast majority were delivered to the Midland lines based at Toton and Cricklewood. During the sixties, the class were a common sight working in pairs on coal traffic, and they helped to displace steam on the Midland. Over the next twenty-five years, they were to became spread far and wide throughout the country.

Until 1979, withdrawals only totalled twenty-seven, but in 1980 fifty-six succumbed and withdrawals continued steadily throughout the eighties until the final twenty-six were withdrawn in the Spring of 1987.

Disposals up to the end of 1986 concentrated on Swindon and Doncaster. However, when Swindon closed in the Spring of 1986, Vic Berry's became involved with the cutting of the final locos. The presence of blue asbestos in some of the cabs cut at Swindon led to them being transported by road to Leicester for processing.

Above: ***In order to create much needed space in the yard during the Autumn of 1987, the cutters finally got to grips with clearing some of the cabs transported to Leicester from Swindon 12 months earlier. On the 3rd. November, cutting of No.25076 was almost complete, whilst behind, Nos.25136 and 46044 were next in line.*** *Photo: Richard Lillie.*

Below: ***Already lifted clear of one bogie, Class 25 No.25060 was being made ready for cutting when photographed on 3rd.April, 1987.*** *Photo: Paul Biggs.*

The first complete 25s to be sold to Vic Berry's arrived in October 1986. Originally consigned to BREL Doncaster for cutting, reorganisation of the works resulted in a cessation of cutting activities at Doncaster. Locos awaiting disposal at the works, including five Class 25s, were towed to Goole on the 14th.August, where they were placed in store. Most were resold to Vic Berry's and arrived at the yard at the end of October.

These five were to be the first of many of the class to arrive at Leicester for cutting during the next twelve months. Not all were immediately cut. So many locos arrived during June and July, 1987, that it was found necessary to strip them of engines and bogies and stack the body shells until such time as they could be cut. Others of the class arriving at the yard remained the property of BR, and were held pending the outcome of negotiations and tenders appertaining to preservation. These were to remain intact and complete in the reception sidings until sold. Vic Berry's then had the task of asbestos removal before these locos could leave for their new homes. By the end of the year, only limited numbers of the class remained in store on BR pending disposal. Once space is available at Vic Berry's these, too, will almost certainly end their days at Leicester.

Cabs sent to Leicester from Swindon Works for asbestos removal.

25076/84/86
25135/36/39/40
25214/36/39/51/56/70/84

Above: ***A close-up view of the cab of No.25154 clearly showing its former number D7504 underneath the paintwork.*** *Photo: Richard Lillie.*

Below: ***Although many Class 25s were placed in the stack and so remained extant at the yard, others were cut soon after arrival. One such example was No.25046 which arrived on the 30th.June, 1987. Cutting had commenced the following day when it was photographed with side panels cut away and components already removed from the engine compartment.*** *Photo: Paul Biggs.*

Class 25 arrival dates.

1986

October

25060/78, 25195, 25257/87

1987

January
25048/64, 25145/75/98,
25200/24/31/69, 25303

February
25051

May
25034/37/95, 25190

June
25027/28/46/75/93,
25123/26/33/34/44/58/60/
61/64/80/93/99,
25207/28/34/44/49/60,
25306/18/27, 25906,
97202(25131)

July
25042/58/59/89,
25153/54/73/91/92/96,
25201/12/30/78/79/85/88,
25323/24, 25902/04/10

September
25057, 25109, 25265, 25313

December
25072

Above: **Withdrawn at Kingmoor in July 1985, where it was stored until March 1987, No.25051 was then moved to Vic Berry's, arriving on the 12th.March. Cutting commenced on the 1st.April, and when seen on the 3rd., only the cabs and bogies remained. The bogies were returned to BR for use on ETHEL 3 97252.** *Photo: Paul Biggs.*

Below: **On 13th.March, 1987, the cutting of former Kingmoor based Class 25 No.25269, was well underway. Note its buffers lying in front of the cab.** *Photo: Paul Biggs.*

Class 25s resold for preservation.

25059	Keighley & Worth Valley Railway
25072	Swindon & Cricklade Railway
25173	Forest of Dean Railway
25191	North Yorkshire Moors Railway
25244	Swanage Railway
25265	Peak Rail
25278	North Yorkshire Moors Railway
25279	Llangollen Railway
25313	Llangollen Railway
25904	Severn Valley Railway

Above: ***Not all the locomotives sent to Vic Berry's have been cut. Several were set aside pending the outcome of bids from preservation societies. No.25278, (unofficially named Castell Powys - Powys Castle), was one of two locos obtained in November for use on the North Yorkshire Moors Railway. Asbestos regulations prevent preservation societies receiving vehicles until all asbestos has been removed. Consequently, No.25278 would need to pass through the asbestos shed before it could make the journey northwards to its new home.*** *Photo: Richard Lillie.*

Below: ***Preparing to depart from Vic Berry's on the 16th.December, 1987, No.25904 is being shunted in the exchange sidings by Leicester-based Class 08 No.08702. It was then tripped to Humberstone Road before being forwarded to the Severn Valley Railway. Put out to tender by BR, this was one of the fortunate Class 25s to be saved by a preservation society. As part of the sale agreement, the society had to pay an additional premium to Vic Berry's for asbestos removal, a pre-requisite before the locomotive could move to its new home.*** *Photo: Richard Lillie.*

Class 26

This was another class to be constructed as a result of the 1955 Modernisation Plan. In the event, it has proved to be a very effective design and remains one of the few original classes still in service.

Designed and built by the Birmingham, Railway and Wagon Company Ltd. at their Smethwick Works, the first twenty members of the class to be delivered in 1958-59 were initially allocated to Hornsey to work local services out of King's Cross. They remained in the Capital for two years before being transferred north of the border to join a further twenty-seven of the class delivered direct. Many were allocated to Inverness and, over the next twenty years, the class became synonymous with the 'Far North' lines. Others of the class were allocated to Haymarket for use in the Fife coalfields. With the arrival of the Class 37/4s at Inverness in 1986, the Class 26s were moved to Eastfield and Haymarket.

Only fourteen of the class have been withdrawn, mostly as a result of accident or fire damage. Of these, eight had been cut at Glasgow Works, four by Vic Berry's at Thornton Yard, and just two had been moved south for disposal at Leicester following cessation of cutting activities by Vic Berry's at Thornton.

An extensive refurbishing programme carried out at Glasgow Works should ensure that the class remains in service for a considerable number of years, and, barring additional accident or fire damage victims, it is not expected that, further examples will arrive at Leicester in the forseeable future.

Above: **Following cutting at Thornton Yard during March 1987, the cabs of No.26033 were transported by road to Leicester for asbestos removal. At the end of the year, at least one of the cabs still remained in the yard awaiting final cutting.** *Photo: Richard Lillie.*

Below: **Only two complete Class 26s have ended their days at Leicester: No.26013 in June 1987, sent from Thornton Junction where cutting had originally been planned to take place, and No.26045 in July 1987 sent from store at Polmadie. No.26013 was seen in the process of being dismantled on the 11th. June, just two days after arrival.** *Photo: Paul Biggs.*

Class 27

As the Class 26s showed much promise when introduced, BRC&W received a further order in 1959 for an additional 69 locomotives, later designated Class 27, to be built with an uprated and modified engine giving an extra 90hp and an increased maximum speed from 75mph to 90mph. Construction was again at Smethwick and the first locomotive was completed in May 1961. The first twenty-three were delivered new to the Scottish Region where they were to remain for their entire working lives, initially on services on the West Highland lines bringing steam services on the route to a close. Of the remaining forty-six locos, nine went new to the North-East before joining the remaining thirty-seven on the Midland Lines based at Cricklewood. Here they remained until the late sixties when they all moved to Scotland to join the rest of the class.

Also of significance in the late sixties was the conversion of twenty-four of the class to push-pull operations for use on an improved Glasgow-Edinburgh service, a duty they performed for ten years before the introduction of the Class 47/7s. In 1979, these twenty-four were converted back, and re-numbered in line with the remainder.

Withdrawals over the years were limited, the first to fall victim being No.D5383 in 1966 following an accident at East Langton, near Market Harborough. Like the Class 26s, the 27s were prone to catching fire, and between 1975 and 1985, a further twenty-two fell by the wayside, a fair number as a result of pyrotechnics. As they were based in Scotland, it was inevitable that most of these withdrawals were scrapped at Glasgow Works although Swindon cut three and Derby, one. Mass withdrawals of the class commenced in 1986 when twenty-five were taken out of traffic, with the remaining twenty-one following in 1987. The vast majority of these were sold to Vic Berry's and during July, August and September cavalcades of 27s heading south to Leicester became a common sight. By the end of the year, only four of the class remained in Scotland awaiting disposal.

During its career, No.27064 has carried four different numbers. Allocated new to Cricklewood in August 1962 when numbered D5409, it was one of the batch converted for the push-pull service between Glasgow and Edinburgh, and renumbered 27116 and later, 27210. When the Class 27s were displaced from this service by the Class 47/7s, it was converted back and given its final number, 27064. It arrived at Leicester on the 17th.August, having travelled south via Mossend Yard, Carlisle and Basford Hall, in company with No.27058. Cutting commenced on the 20th. Note the damage caused to the roof panels when the engine was lifted out. Photo: Richard Lillie.

Class 27 arrival dates.

1987

June
27016

July
27002/04/10/12/45, 27208

August
27017/23/25/26/37/38/41/42/47/48/52/53/55/58/63/64/65

September
27014/22/49/56/66, 27206

October
27030/51

Class 27s resold for preservation.

27056 Northampton Steam Railway
27066 North Norfolk Railway

Two close-up views of Class 27s in the process of disposal. The unidentified cab-front should give anyone modelling a scrap yard excellent ideas and a real challenge to replicate in miniature. Vandals have clearly had a field-day with the control panel and instruments of No.27037. Photos: Richard Lillie.

Class 40

As has been the case with the majority of English Electric products, the Class 40s gave excellent service throughout their lives. Constructed between 1958 and 1962, they first saw service on mainline passenger duties out of Liverpool St., King's Cross and Euston. It was not until later in their careers that they were relegated to lesser turns, often being used on freight traffic. The recession of 1980 and the introduction to traffic of new classes resulted in vast inroads being made into the class. Final withdrawals from capital stock took place in 1985. The works at Swindon, Doncaster and Crewe were responsible for cutting the majority, while Derby and Glasgow both cut one apiece. Vic Berry's did not become involved in the disposal of the Class 40s until after most had already been cut. The five examples to arrive at Leicester originated from Derby and Swindon. At the end of 1987, several of the class were still stored on BR property, pending disposal or the outcome of preservation bids. It remains quite possible that more of the class could be sold to Vic Berry's during 1988.

Class 40 arrival dates.

1987

January
40075, 40132/63

June
40046/63

Cabs sent to Leicester from Swindon Works for asbestos removal.

40015 40149/93

*Above: **Former long term resident at Derby Works, where it had been in store since April 1982, No.40075 was moved to Leicester at the end of January 1987. Cutting was completed within a week of arrival. Behind, fellow stablemates at Derby, Class 45 No.45003 and Class 25 No.25200, await the torch.** Photo: Paul Biggs.*

*Below: **Following its withdrawal from service in February 1983, No.40046 was loaned to the MoDAD at Moreton-on-Lugg for training practices by the SAS. On completion of this duty in July 1986, it was returned to BR and consigned to BREL Doncaster for scrapping. However, cutting ceased at Doncaster before No.40046 could be dealt with and it was sent to Goole for storage, along with the other locos at Doncaster awaiting cutting. Most of these were sold to Vic Berry's but No.40046 again evaded the cutters and was sent to BREL Swindon by way of Bescot and Gloucester. At Swindon it was finally stripped of all internal fittings but, once again, the cutters were denied the task of cutting because Swindon Works closed! Finally, in June 1987, No.40046 was towed from Swindon to Leicester where it was photographed on the 12th. Cutting finally took place later in the month.** Photo: Paul Biggs.*

Class 45

Based on the earlier Class 44s, the 45s were ordered before construction of the 44s was complete. In the event, they proved to be a successful design and have given good service over the years. However, their working lives are coming to an end and 1988 will almost certainly see the final withdrawals. The first example to be withdrawn was No.45067 in July 1977 following severe collision damage. Further withdrawals followed in 1980-81. These early withdrawals were cut at Derby and Swindon Works. Towards the end of 1985, seven of the class were consigned to Doncaster for cutting. However, reorganisation of the Works at Doncaster resulted in them leaving the Works and being sent to Goole for further storage before arriving at Vic Berry's.

During 1986-87, many more of the class were withdrawn and put to store at Toton and March, pending the outcome of studies into the possible conversion of their bogies into snow-ploughs. It may not be long before many of these locos also end their days at Vic Berry's, no more than a stone's-throw from their former haunts on the Midland.

Class 45 arrival dates.

1986

October
45003/23/43/45/55, 45109

November
45016/19/38/42/56/72/75

1987

January
45050/57, 45117

*Above: **A November 1986 view of the reception sidings shows rows of Class 45s and 25s newly arrived at Vic Berry's from Goole and Derby. Present in the view are, on the left, Class 45s Nos.45038/19/43/03, and on the right, Nos.45075, 25287, 46027, 25195, 25078, 25060, 25257.** Photo: Paul Biggs.*

*Below: **Having dealt with No.40075 (see previous page), the cutters moved on to dispose of No.45003. A minimal amount of work remained when photographed on 1st.March, 1987.** Photo: M.John Stretton.*

No.45014 "The Cheshire Regiment" was withdrawn on the 10th.March, 1986 after collision the previous day at Chinley with No.31416. Following storage at Ashburys yard, Manchester, it was moved to the nearby Ashton Road yard where it was cut by Vic Berry's into three sections between the 19th. and 21st.August, and transported back to Leicester by road. The various sections were photographed in the yard on the following day. Photos: Paul Biggs.

Class 46

Ordered in late 1959, construction of this class of fifty-six locomotives took place at Derby Works between 1961 and 1963. Externally, they were similar to the earlier Class 45s, and although equipped with the same Sulzer engines, electrical equipment was supplied by Brush Ltd.

The first withdrawal occurred in December 1977 when No.46005 was taken out of traffic. No.46024 followed in April and No.46003 in October of the following year. All three were cut at Derby Works.

When the recession hit BR in 1980, many of the class were officially 'mothballed' and put to store at Swindon Works. Although a few of the class were later reinstated, most never left Swindon and were cut between 1981 and 1986. Doncaster Works were responsible for the cutting of a further six examples, and should also have cut No.46027. However, like the 45s at Doncaster in the Autumn of 1985, this sole example to arrive complete at the yard had also been sent to Goole for storage.

Class 46 arrival dates.

1986

November
46027

Cabs sent to Leicester from Swindon Works for asbestos removal.

46017/29/44

Although only one complete Class 46 has arrived at Leicester, cabs from several cut at Swindon have entered the yard for final processing. No.1 cab from No.46017 was one of the first ex-Swindon Works cabs to arrive by road for cutting in June 1986. *Photo: Paul Biggs.*

Below: ***The end for No.46044. 3rd.November,1987.*** *Photo: Richard Lillie.*

Electric Locomotives

Class 76

In 1936, with financial help from the government, work began on electrifying the Trans-Pennine route from Manchester to Sheffield via the Woodhead tunnel. During 1941, the first electric locomotive, No.26000, was turned out from Doncaster Works. However, due to the Second World War, all work on the project ceased. Work recommenced on building further locomotives during 1951, this time at Gorton Works, Manchester. No.26057, the last of the class, was completed in 1953. By the late sixties, in common with many British Rail routes, services on the Woodhead route were in decline. Passenger services were concentrated on the Hope Valley line, and Woodhead became a freight-only route.

The first withdrawals of the class took place in March 1970, although no further withdrawals followed until 1977. By then the future of the class, and the Woodhead route, was beginning to look very bleak. The end finally came in July 1981. All remaining members of the class were withdrawn and put to store at Reddish or Guide Bridge. Contracts were issued for their disposal early in 1983, the vast majority being assigned to C.F.Booth's Ltd. Rotherham, with Vic Berry's obtaining three examples which arrived in April 1983.

Class 76 arrival dates.

1983

April
76003/37/40

Class 82

The contract for this small class of ten locomotives was awarded to AEI (Metropolitan-Vickers), who in turn subcontracted the construction to Beyer Peacock of Gorton. Even during the construction stage, it became apparent that the methods used were going to produce a heavy locomotive, and steps were taken to replace some non-load bearing parts with fibre glass. As a result, deliveries between May 1960 and April 1962 were late and protracted.

Two early withdrawals took place as a result of serious fire damage: No.E3055 in September 1969, following a fire three years earlier at Maw Green, and No.E3046 in 1971. Although not dogged with the problems of Classes 83 & 84, the class underwent refurbishment at Doncaster during 1971-72. However, by 1982 most of the class were in store, withdrawal following in 1983. Following storage at various locations, most were finally disposed of at Vic Berry's.

Class 82 arrival dates.

1984

October
82002/04/06/07

1985

February
82001

Class 83

Built by English Electric at their Vulcan Foundry, Newton-le-Willows, the 83s were not EE's most successful design for British Rail. Right from the start in 1960, when the first of the class were delivered, there were problems both with the mercury-arc rectifiers and with unequal weight distribution, giving rise to strain problems on the bogies. So serious were the problems that in the late sixties, BR considered withdrawing the entire class. As a stopgap, pending a final decision, they were placed in store at Bury. As a result of the planned extension of the WCML electrification northwards from Crewe to Glasgow, it was decided that refurbishing the class was a cheaper option than building new locomotives. Accordingly, the class were moved from Bury to Doncaster during 1971-72. However, by 1982, they were again in store, and like the 82s, most were sold for scrap to Vic Berry's.

Class 83 arrival dates.

1984

October
83001/02/05-08/10/11/13/14

Class 84

This class of ten locomotives was constructed during 1960-61 at the Hyde Park Works, Glasgow, of the North British Locomotive Co., under contract from GEC. This was the first class of electric locomotives to be seen running in the Glasgow area, albeit on test trains. Like the 83s, they also suffered serious problems with their mercury-arc rectifiers. In 1962 all were sent to the GEC works at Dukinfield for modification work, but again, like the 83s, found themselves in store at Bury during 1967. Reprieved in 1971-72, they were also refurbished at Doncaster with silicon rectifiers and dual brakes. Unfortunately, new problems came to light and repairs ceased in 1977.

*Right: **4th.November, 1984, and some four weeks after being delivered, Nos.83007/10/02 remain in the reception sidings awaiting their turn to be cut. Externally all look in excellent condition despite having been withdrawn and stored for over twelve months.** Photo: M.John Stretton.*

Withdrawals following quickly. Nos.84003/9 went to the Railway Technical Centre, Derby, where No.84009 was converted to a Mobile Bank Load vehicle for testing ac electrical installations. No.84003 was never converted, and was sold to Vic Berry's in January 1986.

Class 84 arrival date.
1986
January 84003

*Right: **October 1984, and demolition of the LMR electrics is under way. Behind Nos.83005 and 83008, No.82002 has been lifted from its bogies and cutting proceeds.** Photo: Paul Biggs.*

*Below: **Former Manchester-Sheffield electrics Nos.76003 and 76037 are in a slightly more recognisable form than the LT RouteMaster buses which surrounded them. The two Class 76s, along with No.76040, were the first electric locomotives to be sold to Vic Berry's in April 1983, and all were cut within three weeks of arrival.** Photo: Ian Cowley.*

The Atmosphere

Railway scrapyards have a fascination and air all of their own, and have always drawn enthusiasts in large numbers. At no time was this more apparent than during the sixties when private yards proliferated across the country to deal with the 16,000+ steam locomotives being replaced by diesels and electrics. From the large yards of Draper's and Cashmore's which cut many hundreds of locomotives, to the small family businesses whose purchases from BR could be counted on one hand, enthusiasts were drawn like bees to honey pots. At some of the scrapyards, enthusiasts were very welcome, whereas at others they were positively discouraged. What was being cut behind the fence at several yards remained very much a mystery. Perhaps for this reason, steam scrapyards had a special aura about them. Engines would come and go in a matter of days, and even when a yard was monitored on a regular basis, it was not possible to observe the passing of every locomotives.

Through the seventies, BREL Swindon and Doncaster became the new "Meccas" for enthusiasts and they, too, had their own particular atmospheres. With due credit to BREL, enthusiasts were afforded good opportunities to visit these yards, with very few restrictions being imposed.

It was against this background of disposals that Vic Berry's began handling stock. Initially, with only wagons and coaching stock being dealt with, interest was limited to a few dedicated enthusiasts. As the numbers of locomotives have increased at the yard, so has its attraction for enthusiasts. It has to be appreciated that Vic Berry's are a private firm processing scrap metal. To them a Class 27 is just 73 tons of scrap to be cut as profitably as possible. No sentimentality can be attached to stock. It is full credit to the firm that they are able to accommodate limited numbers of enthusiasts. However, because of the yard's geographical location in the centre of Leicester, excellent views can be had from the public footpaths and roads that surround it on all sides. A walk around Vic Berry's is always a sentimental journey. On all sides there is rolling stock that will never run again, and piles of scrap metal, much of it recognisable. Oil, grease, and dirt is everywhere; smoke pervades the scene. This is the modern diesel graveyard, a place of destruction. Yet running concurrent to this theme of decay is the very life of the yard. There is constant movement, from lorries and cranes transporting material, and from the yard's Class 03 ever busy shunting stock. Through the smoky haze can be glimpsed cascades of sparks as the cutters lay yet more stock to rest.

Following page: ***Roger Bamber's dramatic photograph taken in October 1987 brings home the sheer size of this man-made mountain of 25s and 27s. With the photograph on the inside front cover, these pictures are arguably the most powerful and haunting views ever published of the famous stack.*** *Photo: Roger Bamber.*

Above: ***Through the haze of smoke generated by burning oil and grease, sparks from the cutter's torch can just be discerned as more cabs are processed. Behind the coaching stock, a crane is moving more vehicles into position to await disposal.*** *Photo: Richard Lillie.*

Below: ***Purchased from British Rail in 1984 after 25 years' service in the North East, No.03069 sees daily use moving stock around the yard. On a grey winter's morning in December 1987, Class 25s, Nos.25191 and 25278, both destined for a new life on the the North Yorkshire Moors Railway, are moved out of the yard into the exchange sidings to await onward movement to Humberstone Road.*** *Photo: Richard Lillie.*

25
25

27 037

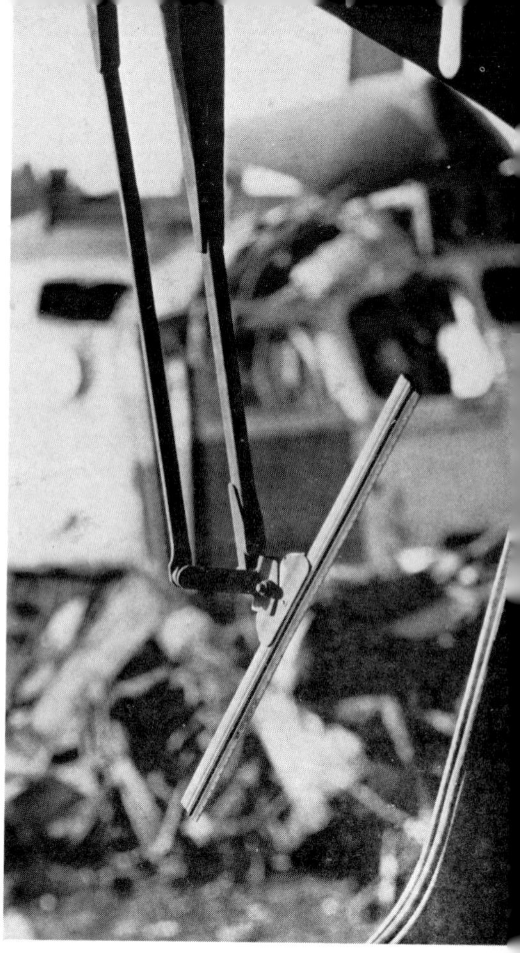

Above: Photo: Richard Lillie.

Top: ***An unidentified Class 26 cab forlornly awaits its fate.*** *Photo: Richard Lillie.*

Centre: ***A visit to the yard always has one wondering ''what will be around the next corner?'' or ''what will be behind that Class 27?''*** *Photo: Richard Lillie.*

Below: ***A wide variety of cabs from Classes 25,26,27,40,45 & 306 still remained at the south end of the yard in December 1987 to await cutting.*** *Photo: Richard Lillie.*

Above: ***Sparks fly from the belly of Class 27 No.27012 as the cutter sets to work removing the fuel tanks.*** *Photo: Richard Lillie.*

Right: ***Surrounded by scrap, one of the cutting team reduces a cab base to manageable pieces of scrap.*** *Photo: Richard Lillie.*

Right: ***"Vic Berry's Merry Men". The cutting team take a breather from work on the Class 20 standing behind them.*** *Photo: Richard Lillie.*

Left: ***Destined never to turn again!***
Photo: Richard Lillie.

Below: ***A 'man of steel', Stephen gets to grips removing bearing metal from a set of wheels.***
Photo: Richard Lillie.

Photos: Richard Lillie.

Diesel Multiple-Units

British Rail are in the process of replacing their ageing fleet of first generation diesel multiple-units, many of which are thirty years old, with a second generation of units. As a consequence, many units in the past few years have been withdrawn from service and passed to scrap merchants for disposal. Second only to Mayer Newman Ltd. at Snailwell, in the numbers of units they have handled, Vic Berry's have, in the four years from 1984-87, processed over 350 units, and look set to handle many more in the coming months.

As with virtually every aspect of their work, it is the need for safe removal of asbestos from within stock that has resulted in Vic Berry's playing such an important role. Units entering the yard first pass through the pre-stripping shed where components such as seats and luggage racks are removed. Vehicles are then transferred to the asbestos shed where, on removal of the internal facia panels, the asbestos sprayed onto the metal body panels during construction is revealed. Once this has been removed, vehicles are towed back into the yard where final cutting is able to take place.

During 1987, Vic Berry's became involved in a programme to remove asbestos from a number of units and return them to BR for an extended life expectancy.

Class 100

Introduced in 1957, this relatively small class were to give BR good service for many years. Withdrawals commenced in 1966, although the majority of the class were taken out of traffic over the period 1971-73. Disposals of these early withdrawals were to a variety of locations: King's at Wymondham, Norfolk, and several yards in Scotland, namely, Ward's at Inverkeithing, McWilliam's at Shettleston, and BR Cowlairs Works. Others of the class survived into the eighties and it was a number of these that were sold to Vic Berry's after being withdrawn from Longsight TMD, Manchester.

Class 100
51127
53351/56/57/58
56105/12/13

Class 101

Built over a period of four years, not only was this numerically the largest class of units constructed, but arguably, one of the most successful. Even by the end of 1987, with many other classes decimated, the Class 101s continue in service around the country. However, as new stock continues to be delivered, their future is looking very bleak and withdrawals are beginning in earnest. It may not be too long before large numbers start to arrive at Leicester for disposal.

Class 101
51196
51502/14
53142
53316
54052
54357/61/77
54403/11
59062/64/67/70/81/87/89/94
59106
59523/29/33/34/35/38/72

Only very limited numbers of the highly successful Metropolitan-Cammell Class 101 units had been withdrawn by the end of 1987, hence their rarity at scrap yards. One of the twenty-seven 101s so far dealt with by Vic Berry's was Chester based No.51196, which was despatched for scrap in May 1986 shortly after withdrawal. By 5th.June, cutting of the stripped body shell had commenced and the job was completed a few days later. Photo: Paul Biggs.

Above: Over the past few years, large numbers of BRC&W Class 104s have ended their working days at Vic Berry's, having been sent to Leicester from the Manchester area. Nos.53471, 53523 & 53436 all came from Newton Heath at the end of May 1986, and were photographed in the reception sidings during the following week. Photo: Paul Biggs.

Class 103

The firm of Park Royal had had a long history of constructing units, having been involved with the pioneer GWR railcars. Surprisingly, when orders were placed in the fifties, Park Royal were only asked to build twenty, two-car sets. Withdrawals of the class commenced in 1971 although it was to be 1983 before the class was extinct. Over half of those cut for scrap have been dealt with at Vic Berry's.

Class 103
50398/99
50400-05/08/09
56150/52/53/55/58/61/64

Class 104

Another successful design, the BRC&W Class 104 low density units were for many years a common sight on the Midland Region on local and semi-fast duties. However, this was not a class chosen for refurbishment in the late seventies and, as new units were introduced, so the 104s were one of the first classes to be withdrawn. Perhaps because they were based on the Midland Region, Vic Berry's have dealt with a high proportion of them.

Below: Cravens Class 105, No.56129, being cut on 1st.May 1987. This particular unit had spent the final years of its life in departmental service as a Sandite Vehicle, No.ADB 977114 before withdrawal in 1987 due to its asbestos content. Photo: Paul Biggs.

Above: During 1987, many of the Class 110 Calder Valley units reached the end of their working lives and, as new replacement units were introduced, so withdrawals increased. As most Neville Hill based units are sent to Snailwell for scrapping, they are not a common sight at Vic Berry's and the appearance of No.51825 in August 1987 was most unexpected. Photo: Paul Biggs.

Class 104
53420/23/28/32/35/36/45/48/49/50/52/57/58/59/61/63/69/71/81/83-86/88/89/90/95/97/98
53502/03/08/09/10/15/23/24/25/27/33/35/37/38/46/51/61/62/65/71/72/90/93/94/95
54175/76/78/80/86/87/89
59139/56/61/65/66/72/86/93
59210/12/14/18/19/26/27/29/40

Class 105
The distinctive front-end design of the Craven Class 105s has been a common sight for many years, although by the end of 1987 few remained in service. Mayer Newman Ltd. at Snailwell, have been responsible for processing the vast majority of the class, while Vic Berry's have only handled limited numbers. Several of these still remain stacked at the entrance to the yard, almost five years after withdrawal.

Class 105
50374/77/94
50757/59/61/94/96
51259/63/71/76/82/94
51491
54417/33/53/61/71/78
56120/36/38/48/49

Class 108
Introduced in 1958, this Derby 'Lightweight' design has proved to be another successful class and featured in the refurbishment programme. Their continued existence in service seems assured for a number of years to come. Withdrawals to date are mostly accident victims and, as yet, only one example has been sold to Vic Berry's for disposal.

Class 108
54200

Class 110
Built in 1961 specifically for use on the Calder Valley services, these three-car BRC&W units are externally very similar to the earlier Class 104s. Only with the arrival of Sprinters at Neville Hill in 1987 have withdrawals begun to make heavy inroads into the class. Few have yet been sold for scrap although a single car arrived at Vic Berry's in June 1987, with a further two vehicles appearing in August.

Class 110
51814/25/44

Class 115
This was yet another class built for a specific suburban service, in this case, working out of Marylebone. Withdrawals to date have been few. In recent years, vehicles have been moved away from Marylebone, notably to Tyseley and it was from there that trailer No.M59718 was withdrawn in May 1984. It was to be a further fourteen months before it arrived at Leicester.

Class 115
59718

Class 116
These three-car high-density suburban units have been a familiar sight in South Wales and Birmingham since they first entered traffic over thirty years ago. Various modifications have been undertaken to the class during their career

and they were included in the refurbishment programme. During 1987, they were finally displaced from many services, either by new units, or older units being cascaded. Most current withdrawals are being sold to Vic Berry's and the class are becoming a familiar sight at the yard.

Class 116
53057/58/62/64/66-69/76/84/88/97/99
53103/04/05/07/10/11/13/18/26/27/30
53821/23/30/31/34/40/45/46/51/52/60/68/75/76/82/83/88/89/92/98
53903/05/12/18
59001/02/05/08/12/13/24
59329/34/39/41/43/50/52/54/60/61/62/66/68/71/73/76

Right: ***As both Tyseley and Cardiff Canton re-organised their DMU fleets during 1987, so the rate of withdrawal of the Class 116 suburban units gathered pace. Still bearing its Welsh Valley emblem, set C301 was in the pre-stripping shed on the 23rd.August, 1987 for removal of all internal fittings. These included the luggage racks which have been stacked in the foreground.*** *Photo: Richard Lillie.*

Below: ***Cutting of the side panels of Class 116 No.53875 resulted in the collapse of the unit due to loss of rigidity imparted by these structures. In the background, a fellow class member has suffered a similar fate.*** *Photo: Richard Lillie.*

Above: ***Only the presence of the condemned symbol painted alongside the BR logo gives away the fact that Class 120 DMBS No.53718 has been withdrawn and awaits disposal. Based at Etches Park, Derby, it had been stored in Toton yard for over a year before it was sold to Vic Berry's in April 1987.*** *Photo: Paul Biggs.*

Class 118

This small class of suburban units was constructed by BRC&W in 1959 to a standard Derby design for use on the Western Region. These too have given good service and withdrawals only began in earnest in 1987. Most of these withdrawn examples remain on the Western Region pending sale to a suitable breakers yard.

Class 118
59476

Class 120

Built to a higher standard than most units, the Swindon Works Class 120s were designed for use on the many long distance cross-country routes that existed in the pre-Beeching era. As lines closed, so the class were transferred around the country. Withdrawals began on a large scale in 1985, and within three years, very few of the class remained in service. Very large numbers of the class have been sold to Vic Berry's both from the Midland and Scottish Regions.

Class 120
51574/76/77/80/85/86/89
51781/89-93
53653/56/57/60/61/63-69/71/73-81/ 83/84/85/89/91/92/94/96/97
53701/02/04/06/07/10/12/14-20/22/ 23/26-29/31/34/35/37/43
59255/56/57/61/63/65/66/67/69/ 72-75/77-80/84/86/87/90/ 91/93-96/99
59300/01
59581/82/83/86/88
59679/80/83/84/85

Class 122

This is a class that has featured frequently at Vic Berry's during 1987, not for cutting, but for asbestos removal, to allow the class to remain in service beyond 1988. So far, only one of the class has been cut at Leicester, and with most refurbished, it is likely that only the departmental examples are liable to early withdrawal and a trip to Vic Berry's for cutting.

Class 122
55002

Below: ***Former Ayr-based Class 120s, Nos.51791 and 51792 and others, unidentified, stacked in June 1986 pending disposal. As 1987 drew to a close, the Class 120s held the record for being numerically the largest Class cut at Vic Berry's. However, it is most likely that they will be overtaken by the Class 116s during 1988.*** *Photo: Paul Biggs.*

Class 123

This was the final class of First Generation DMUs to be constructed. Built at Swindon Works in 1963 for Inter-City duties, they were the most comfortable units introduced. After service on the Western Region, they were later transferred to the Eastern and worked with the Trans-Pennine Class 124 units. Virtually all the class were withdrawn in 1984 and sent to Snailwell for scrapping. Only a single car was sold to Vic Berry's, arriving in January 1985.

Class 123
52098

Class 127

The Class 127s spent virtually their whole lives working the St.Pancras-Bedford commuter services until the line was electrified in 1983. At this stage, many of the trailers were transferred to Tyseley for further use. Of the power cars, twenty-two were converted for parcels traffic, while twenty-one of the remainder were moved the one hundred miles or so from Cricklewood to Leicester where they were quickly cut.

Class 127
51593/95/98/99
51605/07/13/23/26/28/30/34/36/
38/43-48/50

Above: ***A number of Class 122 units have passed through Vic Berry's yard during 1987, not to be cut, but for removal of asbestos, to allow for an extention of the life of the units. No.55006 stands in the new refurbishing shed on 16th.December, 1987, awaiting completion of the work before being returned to BR.*** *Photo: Richard Lillie.*

Below: ***Sole Class 122 to be cut at Vic Berry's is ex-Tyseley unit No.55002 in February 1987.*** *Photo: Paul Biggs.*

Electric Multiple-Units

It is not always appreciated just how many EMUs have been withdrawn from service in recent years, and as new stock continues to be delivered, so the number is constantly increasing. As with DMUs, Vic Berry's have handled numerous EMUs in recent years. Leicester has become one of the very few locations in the country where examples of EMUs from all four electrified regions can be seen together awaiting disposal. The actual disposal process is exactly the same as for DMUs and hauled stock.

It was in 1984, when large numbers of units from around the country arrived together, that the first significant stack was built at the south end of the yard. Composed of units and hauled stock from the Southern, Eastern, and Midland regions, it remained a feature for several months until a lull in arrivals allowed stock to be lifted off and cut.

Class 305

First introduced in 1960, these three-car units were built at York for use on the Liverpool St.-Enfield/Chingford routes. In May 1984, twenty-two sets were taken out of traffic as life-expired and surplus to requirements. Virtually all passed to Snailwell for disposal although two vehicles of set No.542 arrived at Leicester in September 1985.

Class 305
305542 formed 75503/55

Class 306

Introduced in 1949, these were the first electric multiple-units for the Eastern Region's suburban lines out of Liverpool Street. Originally built to operate at 1500V dc, they were converted to 25kV/6.25kV ac in 1960-61. Withdrawals commenced twenty years later in March 1981 and all were out of traffic by the end of the year. Withdrawn sets lingered on the ER at various locations as far apart as Bradford and March, and it was to be 1984 before the final disposals took place.

Class 306
306003*/04*/07/19/20/22*/23*/28/33/34/50/54*/61/67*/71/73/76/77/78*/84*/85/90*
* incomplete sets

Class 309

Introduced in 1962 for use exclusively on the Liverpool St.-Clacton service, withdrawals to date have been limited to the Griddle Cars. They were officially taken out of traffic in August 1982, though they had, in fact, been in store at Wolverton Works for some considerable time previous to this, pending a decision on their future.

Class 309
69100-04/06/07/08

Below: After three years in store at Stanway sidings near Colchester, Class 306 sets Nos.034/61/84/90 all arrived at Leicester in September 1984. They were photographed in the exchange sidings two weeks after arrival. Photo: M.John Stretton.

Class 303

When first introduced in 1959 to the Glasgow suburban services, the "Blue Trains", as they were known, were hailed as the new order for commuter travel. Despite a few initial teething problems, the class have given excellent service for almost thirty years. Withdrawals have been steady over the years and usually as a result of accident damage. However, more withdrawals can be expected in 1988.

Class 303
303017/22/33*/43*/59/84*
* incomplete set

Class 311

In 1967 a further nineteen three-car sets were brought into service in Glasgow. Externally very similar to the 303s, they were built by Craven's and not Pressed Steel who constructed the 303s. Due to the presence of asbestos in some sets, not all the class are expected to remain in service beyond 1988 and further withdrawals are imminent.

Class 311
311100/06

Above: ***One of just two Class 311 sets to have been sent to Leicester, No.311106 was photographed in the reception sidings on 26th.June, 1987. It had been delivered the previous day from Smithy Lye in Glasgow, where it had been in store.*** *Photo: Paul Biggs.*

Below: ***Class 303 unit No.017 had been in store at BRML Glasgow for many months prior to being sold, and this accounts for its derelict state on arrival at Vic Berry's in November 1987. Of the original 91 sets built for the Glasgow surburban lines, 50 have been refurbished and will remain in service for a number of years until new stock can be built to replace them. Withdrawals of the remainder were gathering pace at the end of 1987 and it seems very likely that more of the class will end their days at Leicester.*** *Photo: Richard Lillie.*

*Above: **Although the trailer composites have been withdrawn from the Class 304s, very few complete sets had succumbed by the end of 1987. One of the few victims was Longsight based unit No.031, withdrawn in June 1987 and moved almost immediately to Leicester. The presence of asbestos within this set is evident from the boarding which cocoons the DTBS No.75675.** Photo: Paul Biggs.*

Class 304

In 1960, the LMR introduced the first of forty-five four-car sets built at Wolverton Works for use on the newly electrified Crewe-Manchester/Liverpool routes. Since 1967, only limited withdrawals of the class have occurred, the first being in 1967, following an accident. In 1984, all the trailer composites were withdrawn and the class reduced to three-car sets. All but a few of these TCs were sold to Vic Berry's, the remainder going to Booth's at Rotherham.

> Class 304
>
> Complete units:
> 304018/22/25/31/38/44
>
> TC from following sets:
> 304004/08/10/12/16/17/19/20/21/
> 23/24/27-30/32/33/35/36/
> 37/39-43/45

Class 501

Surburban services from Euston to Watford have been electrified for many years. The Class 501s were built at Eastleigh in 1957 to replace earlier stock and were to give almost thirty years service before they too were replaced. Although Vic Berry's have cut many of the class, not all have been sold for scrap for quite a number remain in departmental service in various guises.

*Below: **The three Class 504s dealt with at Vic Berry's all arrived together in July 1985. As is the case with most units, they were in the yard only for a few weeks before being stripped and cut. By August, DMBS No.65436 had been cleared of all internal fittings and awaited final cutting.** Photo: M.John Stretton.*

Class 501
61135/41/42/47/50/51/52/55-60/64/69/70/71/76/86/88
70135/42/48/50/52/56-61/68/69/71/76/86/88
75148/50/56/58/59/61/88

Class 502

When finally withdrawn in 1980, the last of the Class 502 Liverpool-Southport units wére the oldest EMUs in service, having been introduced in 1939. The two vehicles sold to Vic Berry's had been in store at St.Mary's Goods yard, Derby, for many years, and were the last of the class to be sold for scrap.

Class 502
29557/66

Class 504

The twenty-six Class 504 units were built at Wolverton Works in 1959 specifically for use on the Manchester-Bury route where, with the exception of one vehicle transferred to the Eastern region, they have spent their entire working lives. To date, just over a third of the class have been withdrawn with Vic Berry's handling just three vehicles, the remainder being cut at Wolverton Works or by Bird's at Long Marston.

Class 504
65436/43, 77158

Class 506

Another class built for a specific route, in this case the Manchester-Glossop/Hadfield service, operating at 1500V dc. Following closure of the Woodhead through-route to Sheffield, the Glossop/Hadfield section was converted to 25kV ac and services were taken over by Class 303 sets transferred from Glasgow. The 506s were withdrawn and all, with the exception of set 04 which was saved for preservation, sold to Vic Berry's.

Class 506
59401/02/03/05/06/07/08
59501/02/03/05/06/07/08
59601/02/03/05/06/07/08

Above: ***Delivered new in 1957, the Watford line Class 501 units saw their first withdrawals in 1968 when units Nos.33/34/36/38-40 were taken out of traffic. Further withdrawals followed at a steady rate until the final members were taken out of traffic in October 1985. Many of the final withdrawals were sold to Vic Berry's where, on the 5th.September, 1986, only the driving cab of DMBS No.61170 remained to be cut.*** *Photo: Paul Biggs.*

Below: ***Of the eight Class 506 sets built for the Manchester-Glossop/Hadfield services where they worked for almost thirty years, all but the preserved set have ended their days at Leicester many miles from their former home.*** *Photo: M.John Stretton.*

Above: **The cutter makes short work of the chassis of 4-SUB trailer S9003S from set No.4680. Behind, the body of the vehicle lies on its roof awaiting final demolition.** Photo: Richard Lillie.

Class 405 4-SUB

This numerous class of Southern Region units had its origins back in the pre-grouping days of 1915 when the London, South Western Railway started electric services between Waterloo and Wimbledon. Built for an overhead AC system, these early units were later converted to the now familiar 600V dc 3rd. rail system which became standard on the Southern. In 1939, plans were implemented to build new stock both to strengthen existing sets from three-car to four-car, and also to form new four-car sets. Although none of the original sets survived in service long enough to be sold to Vic Berry's, the firm has handled a number of the later sets. In fact, the first recorded EMUs sold to Vic Berry's included 4-SUBs back in 1977.

Class 405 4-SUB
4279/80/91/94, 4386
4620/33/35*/36/57/60/80/84/87/95
4704*/10/19/33/38/51/54
* incomplete set

Class 410 4-BEP

Introduced to traffic between 1956-63, these four-car sets were built for the Kent electrification scheme. In 1982, when many of the sets were refurbished, the Trailer Buffet cars were withdrawn and sold to Vic Berry's.

Class 410 4-BEP
69000/02-09

Class 414 2-HAP

First introduced in 1954, construction at Eastleigh was not completed until 1963. While many of the class were withdrawn between 1982 and 1984, others were converted both to 4-CAPs and also Driving Motor Brake Luggage Vans for the Gatwick Express. Very many of the withdrawn units were sold to Vic Berry's in 1984.

Class 414 2-HAP
6002/19*/25/26/29/39/47/49/65/69*/76*/79*/91
6107-11/13/14*/15/17/18/22-25/28*/30-33/35/36/37/39/40/43/46/47/48/49*/50/51/52/54/55*/57-61/63/65/67/69
* incomplete set.

Class 415/1 4-EPB

On completion of the final 4-SUB No.4754, Eastleigh immediately commenced construction of another large class of units designated 4-EPB. The first sets appeared in 1951 and over the next five years, 213 sets were built. Many of the class have been refurbished and withdrawals to date are minimal. The first sets arrived at Leicester in 1987 and others, still stored on the SR, could well follow in 1988.

Class 415/1 4-EPB
5127

Cars 14012, 14220/35/51/59, 14334, 14479, 15035/80, 15188 from sets 5006/08, 5110/18/26/30/67/75, 5217/40

Class 415/2 4-EPB

Eastleigh Works was again responsible for this class of seventy, four-car suburban units constructed over the period 1960-63, for services on both the South Eastern and South Western Divisions of the Southern Region. During 1984-85 more than half the class were withdrawn and sold for scrap. Just six sets moved to Vic Berry's, the remainder being dealt with by Mayer Newman's.

Class 415/6 4-EPB
5309/19/38/42/67/70

Class 418/0 2-SAP

This batch of thirty-six, two-car sets were built at Eastleigh in 1958 for the South Eastern Division, utilising underframes from withdrawn 2-NOL units. Just two sets and a single vehicle have been sold to Vic Berry's.

Class 418/0 2-SAP
5601/03*/36
* incomplete set

*Above: **An April 1984 view of the yard shows just three of the many SR Class 414 2-HAP units sold to Vic Berry's.** Photo: Paul Biggs.*

*Below: **An Autumn 1985 view of a coaching stack at the south end of the yard shows the wide variety of EMUs present. Southern Region Class 414 HAP units and hauled stock form the bulk of the stack, while on the top layer, an LMR Class 304 Driving Trailer No.M75686 from set 022 is to the fore. Behind, an unidentified Driving Trailer from ER Class 305 No.542 is also present. To the right of the stack, a Derby Works Class 127 DMU is in the reception sidings. Between the Class 127 and the stack, an unidentified vehicle is being moved ready for lifting.** Photo: Author.*

Departmentals

At the time of Nationalisation in 1948, British Railways inherited a vast array of locomotives and rolling stock of every type and description. One of the first undertakings of the Board was to introduce a new standard type of locomotive and so withdraw from service many of the small, non-standard classes. This idea of standardising the fleet has continued for the last forty years so that, today, BR runs a fleet that shows little variety but makes economic sense with regard to staffing and maintenance. This measure has been of great financial help to the railways although, for the enthusiasts, it has removed much of the interest that went with the hobby forty years ago. A shed visit in 1987 may produce just three or four different classes, whereas the same shed in 1948 may have been home to forty or more classes.

While BRs capital stock may show little variety, its own departmental list is another story. Operated for the engineers and similar departments, this fleet of vehicles is virtually all former capital stock which have been converted for their new role. Many vehicles are old and remain in existence in departmental use long after the remainder of the class have been cut. Consequently, considerable interest has built up in these vehicles over recent years. To preservation societies, they can be a source of rare vehicles, and when tender forms are issued, societies make every effort to secure the most important examples. However, not all withdrawn departmental stock is preserved

Above: One of the more unusual vehicles to turn up at Vic Berry's was CCE Track Machine No.DR 73501 Matisa/Tamper. It had been based in the Sheffield area prior to despatch to Vic Berry's in May 1986. Photo: Paul Biggs.

Below: With the demise of steam heating in coaching stock, the departmental steam heating boiler vans have become redundant and they too are being withdrawn. Nos.ADE 321074 and ADE 321071 were showing signs of vandalism as they awaited final cutting in October 1987. Photo: Paul Biggs.

and scrap dealers have taken many vehicles in recent years, Vic Berry's being no exception. The photographs that follow give an indication of the diversity of vehicles both with regard to their origins and to their departmental usage.

Above: ***Former Craven's DMU parcel unit No.M55997, was converted to Hydraulic Transmission Test Coach RDB 975385 Laboratory Coach No.9 'Hydra' for use at Derby Research Centre. It was withdrawn in 1986 and cutting was almost complete when photographed on the 14th.July of that year.*** *Photo: Paul Biggs.*

Below: ***One of the EMU Class 920 Driving Motors No.ADB 975430, is seen at the yard, stripped and gutted, to await final cutting in February 1987. Originally built as a prototype suburban unit for use on the Southern region, and numbered 2001, the unit was later transferred to Scotland for departmental service.*** *Photo: Paul Biggs.*

Above: ***Departmental RM&EE Office coach No.ADS 70040 stood in the arrival sidings beneath Upperton Road bridge on 22nd.August, 1986. This Southern Region mobile office coach had been based at Haywards Heath but was withdrawn as surplus to requirements.*** *Photo: Paul Biggs.*

Below: ***Another former capital stock coach was BSO No.E9341 which was converted to a Special Instruction Coach, No.TDB 977190 and had been based at Oxley CS. It arrived at Vic Berry's in August 1986 and it was only a matter of days before it, too, became just so many tons of scrap metal.*** *Photo: Paul Biggs.*

Above: ***Former Stratford BTU Tool Van No.ADE 321052, was withdrawn in December 1985 and sold to Bird's Group at Long Marston. However, the discovery of blue asbestos resulted in the vehicle having to be sent to Vic Berry's instead, where it was photographed on 5th.June, 1986.*** *Photo: Paul Biggs.*

Below: ***Not all the stock that arrives at Vic Berry's is actually cut. In use, as a shed for housing machinery parts, is former BTU Staff and Tool Van No.DM 395490. Devoid of bogies, it was photographed on 5th.September, 1986 in company with No.DM 395570, also being put to the same use.*** *Photo: Paul Biggs.*

Hauled Stock

No account of Vic Berry's would be complete without reference to the hauled stock that has passed through the yard.

As is the case with most of BR's locomotives and rolling stock, coaching stock is also being constantly updated and replaced, leaving older stock to be disposed. For the past thirty years, withdrawn coaching stock has been sold to private contractors for disposal. Booth's, McWilliam's, Bird's and Mayer Newman's, to name but a few, are all involved with this disposal. Vic Berry's is no exception, and over the years that the firm have been in business, a constant stream of coaches of every description has passed through the yard. Disposal of coaches is usually rapid, although stacks are usually to be found in the yard as is evident from various photographs in the book. Amongst the coaches which have arrived at the yard are the vehicles from the Colwich crash in 1986. These have been stacked at the yard for many months pending a decision as to their future.

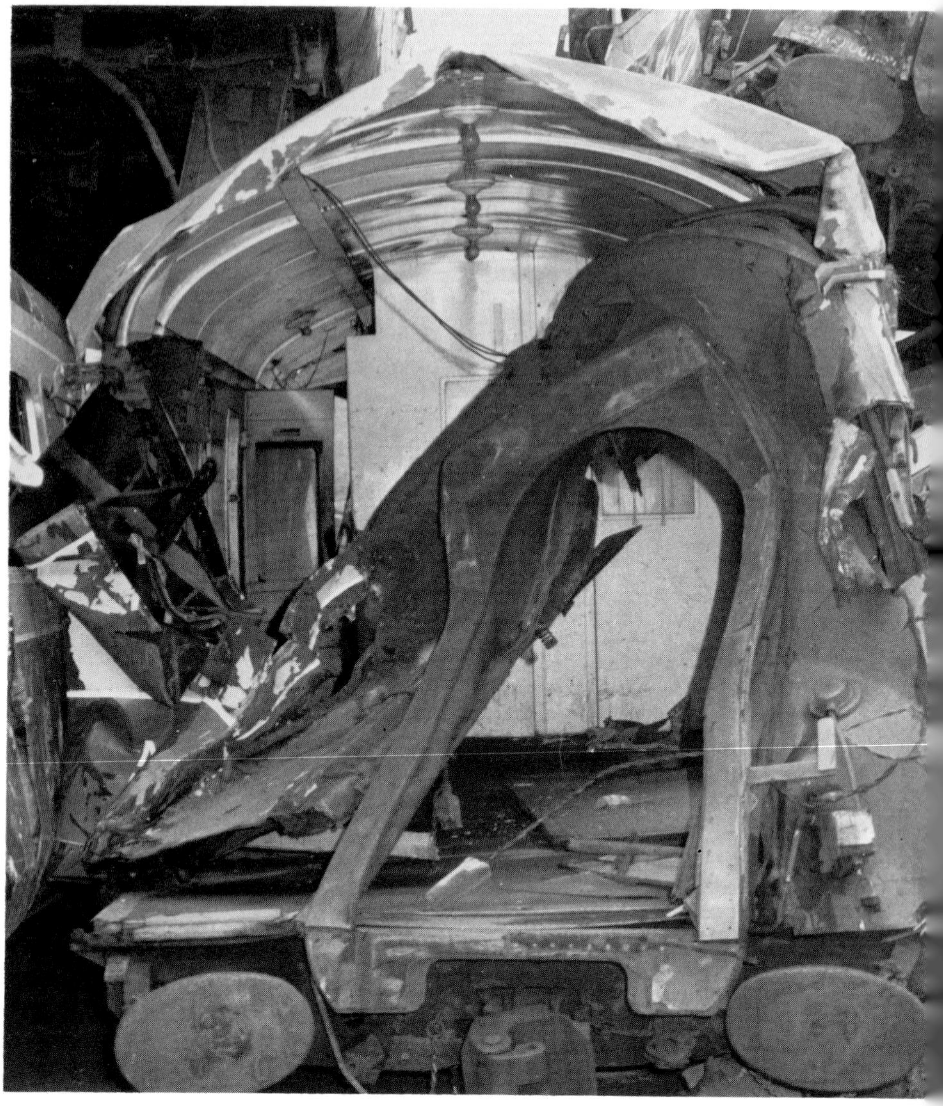

Right: ***The gutted shell of an unidentified coach, photographed in November 1987.*** *Photo: Richard Lillie.*

Below: ***One of just hundreds of similar coaches to be processed at Vic Berry's during recent years is TSO No.M3957. It had been withdrawn from Heaton in February 1986 and stored at York C&W before arriving at the yard in March 1987. It was photographed on the 3rd.April, 1987, with only a few hours remaining before cutting was completed.***

Photo: Paul Biggs.

Above: ***Buffet Restaurant No.M1737 awaits stripping and cutting on the 5th.September,1986.***
Photo: Paul Biggs.

Below: ***The variety of stock sent to Vic Berry's, from locos to units, departmentals to Blue Circle cement Prestflos, is not always appreciated. CPV's Nos.B87370 and B888688 were awaiting further cutting on 5th.August, 1986.*** *Photo: Paul Biggs.*

Photo: Richard Lillie.